volume 4 DEVELOPMENT AND UNDERDEVELOPMENT 1945–1975	volume 5 THE GLOBAL COMMUNITY 1975–2000	volume 6 INTO THE 21ST CENTURY 2000–	chapt	
THE END OF PEASANT CIVILIZATION IN THE WESTERN WORLD	INDUSTRIALIZED AND MULTINATIONAL AGRICULTURE	THE PROBLEM OF SURVIVAL AND THE PROMISE OF TECHNOLOGY	1	Land, agriculture, and nutrition
OLD DISEASES IN THE THIRD WORLD AND NEW ACHIEVEMENTS IN MEDICINE	OVERPOPULATION, DEMOGRAPHIC DECLINE, AND NEW DISEASES	POPULATION, MEDICINE, AND ENVIRONMENT: A RATIONAL UTOPIA	2	Hygiene, medicine, and population
THE AGE OF PRIVACY: HOUSING, CONSUMER GOODS, COMFORT	MEGALOPOLISES IN THE THIRD WORLD; MULTIETHNICITY IN THE WEST	TOO MUCH AND TOO LITTLE: THE MISERY OF WEALTH	3	Living: environment and conditions
WHITE-COLLAR WORKERS, MANAGERS, AND LABOR PROTEST	AUTOMATION AND DECENTRALIZATION IN THE POST-FORD ERA	"THE END OF WORK" AND THE NEW SLAVERIES	4	Labor and production
NUCLEAR ENERGY: THE GREAT FEAR, THE UNCERTAIN HOPE	THE SEARCH FOR ALTERNATIVE ENERGY SOURCES	NEW FRONTIERS IN ENERGY	5	Raw materials and energy
THE PRODUCTION OF THE AUTOMOBILE	ELECTRONICS AND INFORMATION SCIENCE	WORK WITHOUT WALLS	6	Working: environment and conditions
MAN AND GOODS ON FOUR WHEELS	THE AIRPLANE IN MASS SOCIETY	FROM THE EARTH TO THE COSMOS: THE EXPLORATION OF SPACE	7	Transportation
EVERYDAY ENCHANTMENT: THE TELEVISION	THE INFORMATION AGE: COMPUTERS AND CELL PHONES	CYBERSPACE: THE WEB OF WEBS	8	Communication
PARALLEL ROADS: DEVELOPMENT AND UNDERDEVELOPMENT	THE COLLAPSE OF SOCIALISM AND THE RISE OF NEO-CONSERVATISM	THE MANY FACES OF GLOBALIZATION: LOCAL WARFARE AND THE GLOBAL COMMUNITY	9	Economics and politics
MOVEMENTS OF LIBERATION AND PROTEST: THE THIRD WORLD AND THE WEST	FEMINISM, ENVIRONMENTALISM, AND THE CULTURE OF UNIQUENESS	UNIVERSALISM AND FUNDAMENTALISM: THE NEW ABSOLUTISM	10	Social and political movements
CONSUMERISM AND CRITICISM OF THE CONSUMER SOCIETY	THE INDIVIDUAL AND THE COLLECTIVE	AFTER THE MODERN: ENVIRONMENTALISM, PACIFISM, AND BIOETHICS	11	Attitudes and cultures

THE ROAD TO GLOBALIZATION
Technology and Society Since 1800

In private and in public, at work or at play, in every stage of life, we live with technology. It becomes ever more present, and our perception of its artificiality fades through daily use. Within a very short time of their emergence, new possibilities seem to have been with us always, and the new almost immediately becomes indispensable. The choices that technology dictates and the paths that these choices take appear to be the only choices and paths possible—undeniable, unquestionable—and we perceive as natural the constructed world in which we live.

Despite the opportunities that technology affords us, and the promises that it makes constantly, we greet it with a general discomfort, an uneasiness that often does not reach the conscious level. But the manifestations of environmental crises can no longer be considered in isolation. The Westernization of the world marches in step with the widening—and already yawning—chasm between north and south, as well as with the emergence of aggressive localism. War seems to have resumed its role as a common tool in international confrontation. New diseases alarmingly outpace scientific discoveries, and biotechnologies and genetic experiments obscure the line between the human and the inhuman. The importance of the question of meaning has not been lessened by the decline of the sacred; but this question seems to find no place in the universal logic of growth that overcomes difference to guide governing bodies as representatives of economic and financial power.

A renewed uncritical faith in Progress on one hand and a demonization of "techno-science" on the other are often associated with a lack of context that comes of technology and with the dominance of a logic that neglects history. This logic can verify correctness in predetermined ways, but it does not comprehend the complexity of the greater process of change: it appreciates the present and the immediate future, but it cannot perceive itself as part of a larger historical evolution.

The first aim of a social and cultural history of the technology of the last two centuries, then, is to offer a careful and coherent study of the roads that have led to the development of modern culture. The basic objective of this series is the reconsideration of the innovative changes that have taken place and their diffusion over time, rather than a description of their first appearances. These innovative changes have marked and continue to determine our daily lives, the way we work, our relationships, and the points of view that contribute to global diversity.

It is important to recognize that our interpretations of the 19th and 20th centuries are centered on the men and women of the West, on their histories and cultures. This is undoubtedly a biased point of view, and it would be misguided to think that this partiality could be overcome by a simple updating of knowledge. The changing of a point of view that is rooted in history probably requires insight into processes that operate well beyond our perception. Perhaps the globalization that is underway, with its various worldwide effects, is establishing itself through precisely this mechanism: it is forcing a confrontation among lifestyles and different cultural models in new, absolute terms.

In general, the common historiography treats technological innovations only in brief digressions, glossaries, or chronologies of inventions and inventors; but we cannot fill in its gaps by constructing a separate history. Our realization of the economic, social, and cultural importance of industrialization, and our perception of the process as uninterrupted and ever more pervasive, have caused us to re-evaluate both the transformation itself and the new landscapes that industry has created— linking technology to economics and to politics, and systems of labor and production to culture and to social movements.

We can group as the Age of Technology the events that have been paving the road to the future for the last two centuries. Understanding the risks and the opportunities involved in so rapid a transformation of our world will require a change of mind and an updating of our culture—both of which are impossible without a broadening of knowledge and a renewal of historical consciousness.

5

THE GLOBAL COMMUNITY

1975–2000

PIER PAOLO POGGIO
AND
CARLO SIMONI

ILLUSTRATED BY GIORGIO BACCHIN

CHELSEA HOUSE
PUBLISHERS

A Haights Cross Communications ✦ Company

This edition first published in 2003 in the United States of America
by Chelsea House Publishers, a subsidiary of Haights Cross Communications.

Chelsea House Publishers
1974 Sproul Road, Suite 400
Broomall, PA 19008-0914
www.chelseahouse.com

Preceding page: The Minolta 7000, released in 1985, was the first auto-focus reflex camera. Its internal computer first analyzed the framed image, through a sensor built of microscopic light-sensitive elements, and then moved the lens forward or backward until the image was sufficiently clear.

The ancient port structures of Puerto Madero in Buenos Aires, built between 1890 and 1898 to provide Argentina with an international port. The structures were abandoned and for decades remained unused due to conflicts over control of access.

Library of Congress Cataloging-in-Publication Data applied for.

ISBN 0-7910-7096-4

© 2002 Editoriale Jaca Book spa, Milan
All rights reserved.

Original English translation by Karen D. Antonelli, Ph.D.

Cover and design by Ufficio Grafico Jaca Book

Printing and binding by
EuroLitho spa, Cesano Boscone, Milan, Italy

First Printing
1 3 5 7 9 8 6 4 2

INTRODUCTION

Alongside the onset of profound economic and manufacturing transformations, the turbulence and the cultural changes that took place during the crisis of 1968 in developed societies inaugurated a new historic cycle that would be marked by the affirmation of neo-liberalism, by the implosion of the Soviet model of socialism, and by a widening gap between the wealthy and the poor.

In many crucial ways, the closing years of the 20th century were characterized by the overturning of paradigms important to prior decades. In politics, the mass parties and the welfare state entered a period of crisis. In economics, the old system of concentrated and standardized mass production was replaced by a more flexible, personalized production that suited the tastes and the needs of the consumer. There was a simultaneous diffusion of both tired forms of individualism *and* social regroupings that were localistic or artificially neotribal.

A combination of phenomena originating in the accelerated and unbalanced unification of the world, the so-called globalization, resulted from technological advances and the victory of the democratic and liberal system of capitalism, which alone remained standing after defeating or assimilating all competing systems. This political climate evoked a possible, if not favorable, ending to history. Still, new and unforeseen conflicts opened rapidly, beginning with internal crises isolated to individual countries or regions and ending in a series of military conflicts using fewer troops but more technology—and taking a higher toll on the environment.

The 1980s and 1990s saw the rapid spread of electronics and information science, as well as by an exponential increase in the fields of communication, merchandise exchange, and migration. Even with all these internal differences and contradictions, these developments appeared to be uncontrollable and unstoppable. The result was the creation of a dangerous separation between the forms of life in society, which is in constant flux, and the political and institutional order on all levels.

The deep push of the historic industrial cycle continued to produce its effects, turning enterprises and entrepreneurs into the leading figures, the cultural models—the incarnations of the ideals of liberty, competition, and success. In this fashion however, internal inequalities and injustices in a "unified" world became all the more clear. The lack of a means to set social and environmental limits on development aroused insecurity, and fear for the future, in wealthy countries, and a feeling of anger and desperation in the excluded or marginalized nations.

The conversion project of Puerto Madero (1990–1994), an effort to reclaim urban spaces and to recover former port structures for public use, was one of the most successful in Latin America. It enabled the city's people to return to a center that had been in danger of abandonment. Shown here is a three-kilometer stretch of road that runs along four mirror-like water basins; it also fronts a series of old warehouses rebuilt primarily for commercial use but with their original facades intact.

1. INDUSTRIALIZED AND MULTINATIONAL AGRICULTURE

In the final quarter of the 20th century, the industrialization of agriculture reached its acme with the development of two opposing movements that shared the goal of surpassing industrial agriculture. Techno-science offered forms of farming and cattle breeding (or stock farming) that made any relationship with the land appear unnecessary, and biological agriculture viewed the ground as a living organism, not to be polluted by chemical byproducts.

These new scenarios can be understood only in the context of the changes that radically transformed agriculture in both industrialized and developing countries—while the collapse of the Soviet empire quashed its attempts at planning and industrialization in the primary sector.

The salient characteristics of capitalist agriculture at the end of the 1900s were these: (1) an increased output, with the maximal exploitation of the best lands, owing to the massive use of technology; (2) unprecedented development, championed by the United States, of the international trade in agricultural products; and (3) the increased importance of large companies in market production.

Governing bodies were bent on keeping food prices low for residents of the cities; this meant that, even in less developed countries, the farming community was disintegrating rapidly. This disintegration contributed to the massive emigration from the countryside

toward the outskirts of the great cities. Similar results were brought about by the penetration of multinational companies that imposed a single-crop system of farming, even through military control. (One example of this forced specialization in farming now feeds the wealthy international drug market.) Notwithstanding the ruling climate of neo–free trade or *laissez-faire*, in no other economic sector can we see such an intense intervention of nations and

1. *Coffee-bean pickers in Kenya. Typical plantation farming: for a long time the cultivation of coffee had been under the control of white colonials. Following its independence, Kenya continued to depend on the export of tea and coffee, the prices of which underwent dramatic swings tied to the international markets.*

international organizations by means of agreements, duties, quotas, regulations, and tariffs. These interventions were unable to avoid instability in prices or to generate surplus production. Instead, they led to a scarcity that perpetuated hunger.

Industrial agriculture was a great success in the short term, but it also led to troubling situations like erosion, a lowering of the water table, and the transformation of fertile land into desert. But industrial agriculture negatively affected more than just the land: the use of animal-affected flour caused bovine spongiform encephalopathy (BSE, or "mad cow disease"), leading to one of the major scares of our time.

Early in the 1970s, agriculture began to feel the impact of developments in genetic engineering, such as the production of genetically altered plants—a production dominated by multinational nutritional, chemical, and pharmaceutical companies.

2. Malaysian palm-oil plantations in the 1970s. In an attempt to decrease or eliminate its dependence on rubber— 30% of its exported product—Malaysia strongly encouraged the growing of oil palms. The amount of cultivated land doubled between 1972 and 1975.
3. "Container" ship. The 1960s and 1970s saw the establishment of an interdependent worldwide agricultural system using major navigation routes, as well as cargo planes for perishable goods.

2. OVERPOPULATION, DEMOGRAPHIC DECLINE, AND NEW DISEASES

At the beginning of the 19th century, there were 1 billion people on the Earth. By 1959, the number had tripled, and in the following 40 years that number doubled again. In 2000, there were almost 6 billion people in the world, and every minute some 150 more were born. Most of these births took place in the 122 countries of the Third World, where having several or many children is believed necessary to guarantee, if not economic success, at least survival. In the countries of the northern part of the world, on the other hand, the birth rate had been dropping for many years, and without the demographic contributions of immigration,

the populations of some European nations would actually have declined.

Notwithstanding this phenomenon of "zero population growth" in wealthier countries, the world's population increased by about 80 million people every year, an increase limited considerably by the effects of malnutrition. Some 500,000 African women died in childbirth yearly owing to the effects of

malnutrition; malnutrition also led to diseases unknown or forgotten in the developed world, like *kwashiorkor*, a protein deficiency that slowly destroys the bodies of undernourished children, or the intestinal diseases that attacked

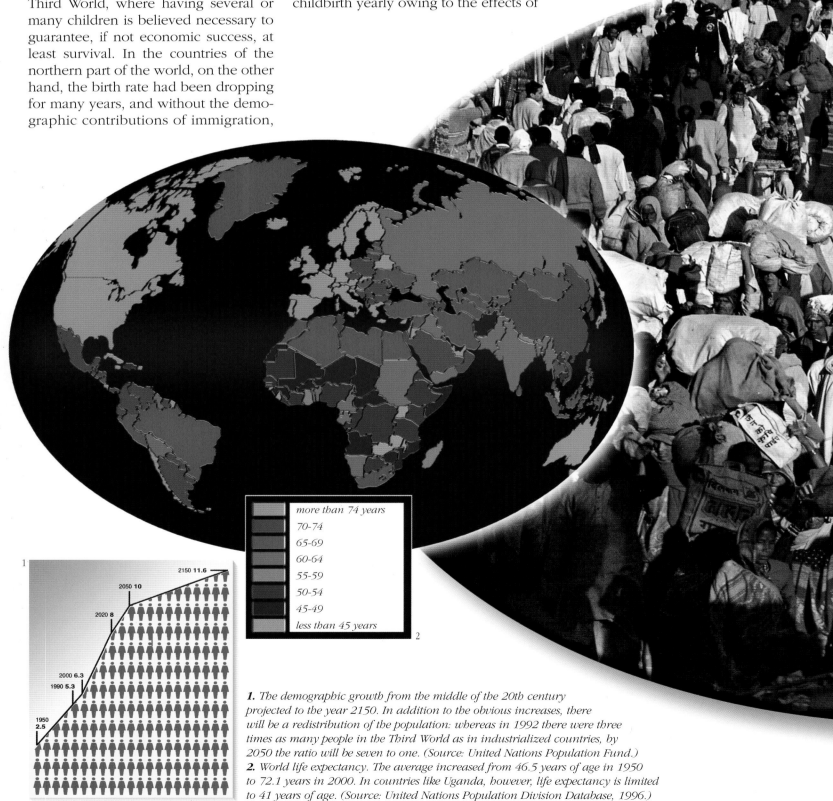

more than 74 years
70-74
65-69
60-64
55-59
50-54
45-49
less than 45 years

2150 **11.6**
2050 **10**
2020 **8**
2000 **6.3**
1990 **5.3**
1950 **2.5**

1. The demographic growth from the middle of the 20th century projected to the year 2150. In addition to the obvious increases, there will be a redistribution of the population: whereas in 1992 there were three times as many people in the Third World as in industrialized countries, by 2050 the ratio will be seven to one. (Source: United Nations Population Fund.)
2. World life expectancy. The average increased from 46.5 years of age in 1950 to 72.1 years in 2000. In countries like Uganda, however, life expectancy is limited to 41 years of age. (Source: United Nations Population Division Database, 1996.)

millions of inhabitants of shantytowns, who searched for food in the mountains of garbage produced by the wealthier classes in the cities near which they lived.

While silence surrounded *these* deaths, huge newspaper campaigns accompanied the epidemics that spread through the northern part of the world. BSE was even more unsettling because it was transmitted to humans not by a virus or bacteria, but through the proteins contained in meat, and because it was found predominantly in the countries that on average consumed more meat. In these countries, meat was seen as the most nutritious food, a symbol of health and wealth. Moreover, the incubation period of BSE was not known, and this fact above all others likens BSE to Autoimmune Deficiency Syndrome (AIDS), the rise of which shattered the illusion of victory against infectious diseases. The first case of AIDS was documented in Congo in 1959, but the disease became widespread only 20 years later. It did spread throughout the countries of the West, but it hit Africa especially hard. Out of a worldwide infected population of 36 million, nearly 28 million were found in sub-Saharan Africa. In the Republic of South Africa, 1,600 new patients were discovered to be HIV-positive every day. Two hundred of these were newborns, and 25% of these infants would not reach their second year of life.

3. The enormous gathering on the occasion of Kumbha Mela, the largest Hindu pilgrimage of all time. It was held in 2001 at Allahabad on the Ganges River in India. In 2040, India will have more inhabitants than China, and the two nations together will be home to one third of the world's people. (Photograph by Matteo Rodella.)

4 and 5. One child, two children at most, is often the norm in Western families. (Photograph by W. Boke/Berg/Image Bank.) A decrease in the number of births and an aging of the population— here we see two elderly people seated on a bench near the Bay of Montevideo—are the two related phenomena that characterize the current demographic trend in industrialized countries.
6 and 7. Left: The Ebola virus, which is responsible for a high mortality rate in Africa. Right: The antibodies that react to the HIV virus under a microscope. (Photographs by Marka.)

3. MEGALOPOLISES IN THE THIRD WORLD; MULTIETHNICITY IN THE WEST

In the last decades of the 20th century, the configurations of the great cities of the world changed significantly, in both industrialized and non-industrialized countries. This change resulted from urban expansion and increasing immigration from the countryside and led to the formation of megalopolises, enormous urban areas that, although similar in size, were profoundly different from one another. In the West the additional growth of the metropolises was due to the shifting of production activities and to the housing of the middle and upper classes in different suburban areas. In poor countries, on the other hand, the suburbs were populated by millions of individuals who had abandoned rural areas. Shapeless and frightening communities, in which people lived in hellish conditions, surrounded Mexico City, Cairo, São Paolo, Buenos Aires, Bombay, Tehran, and other cities.

The megalopolises of the Third World, which expanded like drops of oil on water, generally lacked infrastructures and public services and were socially precariousness, with marginalized groups fighting daily for survival. The continual growth of these cities can be attributed to the catastrophe of the farming world and was fed by a failing politics of development and by the demands of the dominant model of production, which was built on the availability of cheap labor.

The causes and the dynamics of this varied considerably, but the global tendency was a rapid and continuous growth of urbanization: in 1950, only 29% of the world's population lived in cities, and by the end of the century the proportion had reached nearly 50%. In wealthy countries, even beyond the demographic reductions, the phenomenon of a "return to the countryside" had been in place for some time, and

the increase in urbanism was found mainly in underdeveloped or developing countries, especially in Asia. In the 1990s, 14 out of 20 cities whose populations surpassed 8 million were located in developing countries.

But the migratory movements were not directed only toward the vast outskirts of these cities. Beginning in the 1980s, there was a significant increase

1. Mexico City. Built on the site of Tenochtitlán, an Aztec city destroyed by the Spanish, Mexico City is the Third World megalopolis that saw the fastest and most chaotic growth in the 20th century. From 350,000 inhabitants, it reached 11 million by 1975 and over 20 million by the end of the century. (Photograph by Angelo Stabin.)

in the number of migrants who headed toward the cities themselves and toward the industrial areas of the West. This increase was not limited to the United States; it included Western Europe. For the U.S. this was a continuation and reaffirmation of what was already in place, but in Europe multiethnic cities raised many new problems, linked to different interests, cultures, and values. Many were ready to reap the rewards of interaction with others, but there also was rejection; some attempted to revive racist or xenophobic political ideologies.

2. Murals in San Diego, California. Chicanos (Mexicans) reclaimed their place in American society, reminding the nation that their "ethnic group" was no longer a minority in that part of the country. (Photograph by Promotion & Visuals.)

3, 4, and 5. Boston. The capital of Massachusetts, one of the oldest American cities and an intellectual and academic center of the highest importance, is also characterized by the presence of a broad ethnocultural stratification: from the Anglo-Saxons, descendants of the Puritans, to the Irish and the Asians. (Photograph by Angelo Stabin.)

4. AUTOMATION AND DECENTRALIZATION IN THE POST-FORD ERA

In the years following World War II, the availability of feedback mechanisms hastened the spread of technical automation, which many saw as a path to an ever-expanding replacement of humans by intelligent machines. Automated systems spread widely throughout production, for example in machine tools with numeric controls, and in the technical devices used in every aspect of daily life. Still, automation did not

of computers and computer science—from microprocessors to the widespread use of personal computers and their connections to the Internet—for exponential growth.

On this technological base a deep transformation of the production processes and of international division of labor developed. Manufacturing ventures decentralized ever more substantial portions of production through

controls. This meant movement to Eastern Europe for the Western European businesses and to Mexico for the American businesses.

The post-Ford era strongly reaffirmed capitalism and a reversal in the workers' movement, which was forced to accept the order of the day with flexibility. The key elements of the new production model were the financing of the economy, the control

spread to the entire productive process; it remained fragmented. The fully automated large factory had (and has) yet to be built.

The Third Industrial Revolution, known as the post-Ford era, followed other roads, taking advantage of automation and robotics whenever convenient and, above all, of the potential

a myriad of suppliers and subcontractors, keeping "in house" only the design and overall management and at times only the trademark or logo. In many cases, this outsourcing translated into a movement towards countries with more favorable economic and environmental situations—in other words, lower salaries and fewer

of the flow of information, and the logistics of resources, including the labor force, on a global scale. Beginning with the 1980s, the paradigm of the post-Ford era, having established itself in the manufacturing sector, extended throughout society with the privatization of the public sector.

1. *Teresa Maresca, Fabbrica (Factory), 1999 diptych. Technological and organizational change was a strong incentive to abandon antiquated industrial buildings; the abandonment created a landscape of forsaken factories that attracted artistic attention and raised the problem of relating to the memory of industrialization.*

2. *The automated sector of a textile factory. Manual labor disappeared from many phases of the production cycle, especially in older areas like those of spinning and weaving, which were traditionally characterized by the presence of female workers.*
3. *Trucks used for the transportation of various types of merchandise. The organization of labor in the post-Ford era, founded on decentralization and networking, greatly increased the movement of raw materials and of semi-finished and finished goods.*
4. *Workers in a garment factory in Seoul, South Korea. Multinational enterprises in this sector have definitively focused on several Asian countries, like South Korea, drawn by the competitive advantages they can derive from low wages and a lack of unionization.*

5. THE SEARCH FOR ALTERNATIVE ENERGY SOURCES

The energy issue, which was closely linked to environmental issues, was the backdrop for the historic events of the last decades of the 20th century, and it is essential to their interpretation. The predictions of a depletion of the sources of petroleum, and above all the two crises of 1973–74 and 1979–80, forced the industrialized countries to search for alternative sources of energy and for ways to conserve energy. Further cause for reform came from the recognition that atmospheric pollution resulted from the use of fossil fuels—and by the greenhouse effect, caused by carbon dioxide in the atmosphere, which altered climates in increasingly noticeable ways. From the 1970s onward, numerous policies were initiated to confront the energy crisis and to encourage both conservation and the development of alternative sources.

The International Energy Agency (IEA) was established in 1974 to spearhead this movement, but it failed to accomplish what it needed to accomplish and what had been expected of it. This was for two major reasons. First, despite the idea of sustainable development, there seemed to be an irresoluble contradiction between the dominant industrial model and a new paradigm, which would have required lifestyle changes as well as major technical break-throughs made independent of market forces. Second, powerful economic and political interests managed to keep the price of petroleum low; this slowed the search for alternatives.

At first, the answer to the petroleum crisis was sought in nuclear energy, but the disastrous nuclear meltdowns at Three Mile Island (1979) and Chernobyl (1986) discouraged most countries from continuing along this path. Attempts to obtain energy from nuclear fission, including the so-called "cold fusion," remained experimental.

The search for alternative energy produced important results, if not always *favorable* results, in all the major sectors of renewable resources—wind energy, based on perfecting an ancient technology; solar energy, especially with the use of photovoltaic cells; electric power from the oceans, obtained by a complicated process of thermal exploitation such as Ocean Thermal Energy Conversion (OTEC); and biomass energy, which is not entirely safe for the environment.

1

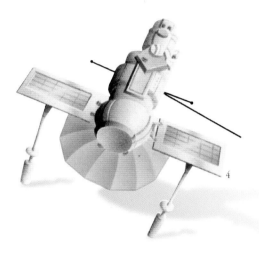

1. A gigantic pipeline through a shantytown in the outskirts of a Third World city. The increasing inequality of access to resources between rich and poor is the most remarkable facet of recent history.
2. High tide and low tide at Saint Malo, the location of a tidal power plant on the Rance River in Brittany, France. Harnessing the energy of the tides, possible only under certain conditions, is among the alternative energy sources under consideration. France was among the first countries to explore this option, with the tidal power plant built on the estuary of the Rance River. (Photograph by Hervé Boulé/Edilarge; map by Ermanno Leso.)
3. Geothermal electrical plant. The endogenous underground heat represents one of the most interesting primary sources of energy. Exploited for a long time in Iceland (geysers) and in Italy (the Larderello plant in Tuscany), this form of energy has inspired many studies and much research aimed at identifying new geothermal areas. (Photograph by W. Higs/GSF Picture Library, published in "Earthwatch: Fuelling for the Future," A&C Black, London.)
4. The Soviet space probe Venera 4, fueled by solar panels. (Illustration by Roberto Simoni.)

6. ELECTRONICS AND INFORMATION SCIENCE

The electronics industry, in particular micro-electronics, has been characterized by an impetuous growth. Beginning in the 1970s, this growth was enhanced by the industry's strong relationship with computer science. Originally, the United States occupied the dominant role; in 1975, it produced 64% of all the world's semiconductors. Several Asian countries then established themselves: the birth of the commercial software industry, the field that established the era of Silicon Valley.

Electronics and information science held a strategic role in all contemporary economics, and their incidence on a social level goes back to a solid anchoring in the manufacturing field (from the planning stage to applied robotics). Micro-electronic and information-science

Japan first, then Singapore, Malaysia, Korea, and Taiwan, and finally the giants India and China. The American position weakened, but only in the industry of hardware, the machines themselves; the situation was better in software, the programs that operated the hardware. The availability of personal computers brought about

technologies spread through every sector of activity; in the first place inside companies, profoundly changing their functions and contributing to the revolution in the organization of labor that had first developed in the 1970s. The primary strategies of the new productive model—from the *just in time* model, which refers to the elimination of the

1. A toy robot. The word robot *was popularized by the Czechoslovakian writer Karel Capek in his play* R.U.R. *(* Rossum's Universal Robots *), written in 1920 and translated into English in 1923, to indicate mechanical devices in human form that worked in the place of laborers. The first robot controlled by a minicomputer was put into commercial use in 1974 in the United States, launching the robotics industry.*

warehouse and of all dead time, to the *outsourcing* model, or the externalizing of work—were made possible by the real-time control of all of the moments of the work cycle, from production to consumption. The projection of the factory in space and the interconnectedness of the many items worked on was made possible by the continuous spreading of information networks that tied together machines, the workforce, and tangible and intangible resources.

2. The integrated circuit board of a microprocessor. The development of the electronics and information-science industries was accelerated by the invention of the microprocessor, with its ever-expanding memory, at the end of the 1970s. (Photograph by David McGlynn/Marka.)

3. Digging in a gold mine in the State of Pará in Brazil. Gold is one of the raw materials used in microelectronics, especially for the welding of circuits. Countries like China are specializing in the important activity of salvaging the precious metals used in electronics. (Photograph by Mireille Vautier.)

4. A robot in action on a Fiat assembly line. The introduction of the robot has radically changed the work in large automobile factories.

7. THE AIRPLANE IN MASS SOCIETY

A very limited number of passengers, gathered in a cold and noisy space, uncertain of the safety of the means of transportation, forced to endure numerous interruptions at the layovers needed for frequent refueling: this is an image of the early days of travel on commercial airplanes, usually biplanes whose aerodynamic wing resistance limited both their flight speed and altitude. Beginning in the 1930s, planes

based on personal perceptions to a purely technical experience, in which the instruments that filled the cockpit replaced the reference points of the sky and the ground below. This evolution extended also to passengers, as can be seen in the advertisements of the airline companies, which ceased to advertise the possibility of enjoying spectacular scenery and instead emphasized speed, safety, and above all comfort.

overcame many of these limitations; they were constructed entirely of metal and their cabins pressurized. Also, many of the technologies developed in World War II—in which the use of massive air bombings, for example, was a dramatic innovation—were transferred to passenger planes.

The first jet plane, the *De Havilland Comet*, entered commercial service in 1952. By the early 1970s, the gigantic "jumbo jet," capable of transporting a large number of passengers, had

opened the era of mass flight and the Russian *Tupolev Tu-144* and the Anglo-French *Concorde* had inaugurated civilian flight at speeds faster than that of sound. In June of 1993, an Airbus A340 demonstrated its range and fuel capacity by circling the globe with only one stop.

On a course parallel to these radical and very rapid developments, the significance of flight and the ways in which it took place also changed. The role of the pilot passed from a flying

Even airports, in their ever more homogeneous and functional architecture, seemed to diminish the experience of flight once it had come into mass use. Still, for many passengers flying was a source of curiosity or even fear. Airports transferred passengers from waiting rooms to airplanes by buses or moving sidewalks, conveying them from the armchairs in which they had awaited departure to the armchairs in which they would await arrival. Beyond the disruptions due to accidents

1. *The Wright Brothers, Wilbur (1867–1912) (shown in the photo) and Orville (1871–1948), who ran a small bicycle factory, planned the first motorized airplane. The first successful experiment in flight occurred in 1903: their biplane flew some 36 meters at a height of 3 meters above the ground. (Photograph by Image Bank.)*
2. *A modern civilian airplane. The flight takes place above the clouds and atmospheric disturbances, avoiding the dangerous turbulence of lower altitudes. (Photograph by Marka.)*
3. *The Flying Wing, a design for a civilian airplane now only in the planning stages, is very similar to the design of the Stealth bomber. "Flying wing" aircraft will be able to carry about 800 passengers in the space inside its wings. (Illustration by Ermanno Leso.)*
4. *A cargo plane in action. The size of airports and the high volume of passengers that travel through them have created a need for machines that speed up the operations that precede takeoff and follow landing. (Photograph courtesy of Ali Viaggi.)*
5. *A plan of the spaces and functions of the Charles de Gaulle airport in Paris, France, which employs over 60,000 workers. (Illustration conceived by Bruno Le Normand, realized by JSI, courtesy of Air France.)*

A. Security check and/or passport control
B. Customs area
1. Information desk
2. Ticket sales
3. Check-in counters
5. Waiting lounge
6. Connecting-flights counter
7. Baggage service
8. Bus stop

and re-routings, for those who used airplanes not for business reasons but for pleasure and tourism—especially after September 11, 2001—doubts arose similar to those that had accompanied the appearance of the train. Many feared that travel, executed ever more rapidly and with ever less contact with the environment and the scenery, was in danger of losing most of its significance and would end in spiritual impoverishment.

6, 7, *and* **8.** *Three images of daily life in airports: Logan in Boston, Charles de Gaulle in Paris, and the airport of Kathmandu, Nepal. The complex system of work spaces, shops, restaurants, and waiting rooms that spread out from around the runways turns airports into virtual cities, crowded at any hour and largely self-sufficient. (Upper and central photographs by Angelo Stabin; lower photograph courtesy of Ali Viaggi.)*

8. THE INFORMATION AGE: COMPUTERS AND CELL PHONES

Norbert Wiener (1894–1964), one of the fathers of cybernetics—the development of machines capable of reproducing the functions of the human brain—claimed that having surpassed the physical strength of human beings, machines were on the way to surpassing the human intellect. This would have been possible thanks to the development of information technology, an exemplary case of the meeting of science and technology, which joined mathematics, physics, chemistry, and the methods of processing silicone. In the 1960s, a way was found to embed thousands of transistors into small amounts of this element (silicon chips). These devices then took the place of cumbersome valves (electronic tubes), which, in addition to radios and televisions, had also been used in the first computers.

1. A detail of the computer EDSAC (Electronic Delay Storage Automatic Computer), designed and built at England's Cambridge University in 1949. Through the use of tubes, it became the first completely electronic calculator equipped with a memory function.

At first these computers were so large that one filled an entire room; but by the end of the 20th century laptop computers were as small as a book. Prices have decreased, too, so much so that if the same trend had been registered in the cost of passenger airplanes, by the end of the century about $500 would have been enough for the purchase of a DC-9. The power and memory of computers—as well as their maximal operating speed, which doubles every 18 months—continue to *in*crease.

Equally dizzying has been the speed with which personal computers proliferated in the workplace and at home, becoming an everyday presence for millions of people. No less surprising is the computer's invasion of other sectors; it became the essential base, for example, in the areas of biotechnology,

transportation systems, and communications. In this last field, the automation of the treatment of information is tied to the automation of its transmission.

Still in the communications sector, another instrument spread with an extraordinary speed and ability to change habits, lifestyles, and modes of communication and interaction. At the beginning of the 1990s, the cellular phone was considered by many an instrument of little use, certainly not indispensable, and therefore destined for use by an extravagant elite. By 2000, though, it had become a means of communication that—despite the apparent harmful effects of its use—a growing number of people would have been reluctant to give up. A "personal" object in the same vein as portable radios and CD players, the cell phone could accompany its owner anywhere; this engendered a feeling of immediate access to communication. The phenomenon seems irreversible, even though 84% of users live in industrialized countries and more than two billion of the world's people have never made a telephone call.

2. A silicon crystal. After oxygen, silicon is the most common element in the Earth's crust, representing nearly 26% of its weight. It is found predominantly in silica, a substance available in quartz and other forms. Silicon is the basic element of microelectronics.
3 and 4. Spending hours in front of the computer screen has become a necessity in many jobs, both in industry and in service, but the computer can also be used for education and entertainment. (Upper photograph by Henry Sims/Image Bank; lower photograph by Tom Stewart/Contrasto.)
5. The information-science and telecommunications industries hire millions of people. In the last few years, they have been a major force in the financial world and have led an increase in production. (Photograph by Brownie Harris/Contrasto.)
6. The cellular telephone, not merely an indispensable tool for millions, has become an object of design. Efforts are constantly being made to decrease its size and weight.
7. Walking alone while speaking animatedly is an action spread by the cellular phone which a few years ago would have caused a reaction in others. (Photograph by Angelo Stabin.)

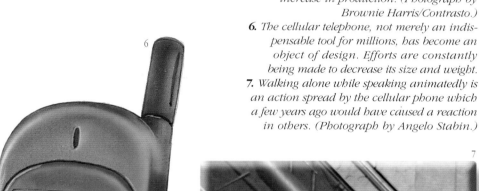

6

5

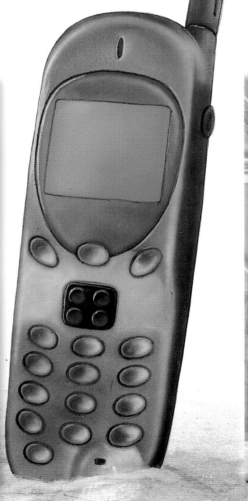

7

9. THE COLLAPSE OF SOCIALISM AND THE RISE OF NEO-CONSERVATISM

The USSR's intervention in Afghanistan in 1979 marked its moment of maximum expansion in the world and the beginning of a decline that soon became unstoppable. Internal dissension had appeared and gained strength through dissidents like the novelist Alexander Solzhenitsyn and the physicist Andrei Sakharov, but was not echoed on the social level. The first important breakthrough took place in Poland through the actions of the union movement Solidarnosc (Solidarity) and with the influence of the Catholic Church, whose point of reference was

Union of Soviet Socialist Republics—in reality satellite countries dependent politically and economically on the USSR—manifested itself painlessly, like the crumbling of an empty shell. In fact, the collapse was like a story that had already ended.

A situation similar to the collapse of the satellite block occurred also *within* the USSR, despite the fact that it was a military superpower and had overcome

system, but he was not successful until an attempted neo-communist coup in August of 1991, which became the springboard for Boris Yeltsin's rise to power in December of 1991 when the USSR formally ceased to exist.

At the same time, other events brought an end to "realized socialism": a strong neo-conservative wave developed in the West, initiated in Great Britain by Prime Minister Margaret

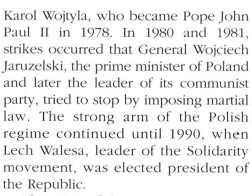

Karol Wojtyla, who became Pope John Paul II in 1978. In 1980 and 1981, strikes occurred that General Wojciech Jaruzelski, the prime minister of Poland and later the leader of its communist party, tried to stop by imposing martial law. The strong arm of the Polish regime continued until 1990, when Lech Walesa, leader of the Solidarity movement, was elected president of the Republic.

The crisis of the Soviet system was exacerbated by the collapse of the Berlin Wall in November of 1989. With the exception of Romania, the end of the

crises of extreme gravity before. The most common explanation was that the Soviets had sapped their resources in their rivalry with the West, losing ground mainly in the economic arena. The USSR was completely unprepared, despite a strong technological and scientific background, to confront the mandates of the Third Industrial Revolution, the revolution in electronics and information. From 1985 onward, Mikhail Gorbachev tried to reform the

1. Monument to "Mother Russia" near Volgograd (formerly Stalingrad). The entire story of Russian socialism was marked by the exaltation of patriotic and military themes, touted against fascism and anti-capitalism. (Photograph by Cornelius Meffert/Stern.)
2. Female Russian workers in an asbestos factory in 1993. Environmental concerns and health questions regarding the workforce were never a priority in the former Soviet Union, and the situation has not improved under the new regime. (Photograph by Peter Turnley/Grazia Neri.)

Thatcher in 1979 and affirmed by the election of Ronald Reagan as president of the United States in 1980. It was not merely a movement toward the right by the majority of the electorate, but a much deeper phenomenon that reflected social and economic transformations. It is not by chance that in all the developed countries the attack against governmental interference and against the excessive costs of welfare,

that is, the fight for privatization, flexibility, and competition, became the platform of the entire political class, which tried in this way to keep up with the times. This stripped politics and

democracy of meaning, and citizens retreated into privacy or searched for alternative ways to face the growing problems that resulted from the dynamics of the later industrial societies.

3. Solidarnosc (Solidarity): This organization, led by Lech Walesa, was the first great mass movement—after the revolts of the 1950s—that arose in opposition to a communist regime in Eastern Europe.
4. Debris from the Berlin Wall at Pankow in East Berlin in 1990. The destruction of the wall that divided the German capital was an event of enormous symbolic importance and the prelude to the rapid reunification of Germany. (Photograph by P. Habans/Sygma/Grazia Neri.)

5. Lech Walesa and U.S. President George H.W. Bush in Danzig in 1991. The United States backed the Polish leader's free trade–based and authoritarian politics, which inspired varied criticism, even within Solidarnosc.
6. Coca-Cola in the Imperial Palace in Beijing. After the death of Mao Tse-tung in 1976, China began a process of rapid modernization. (Photograph by Liu Heung Shing.)
7. The GUM superstores in Moscow. Among the West's most powerful weapons in the face-off with the Soviet system was a marked difference in levels of consumption. Mass communication revealed the difference, proposing an alternative model that took hold easily. (Photograph by D. J. Heaton/Marka.)

10. FEMINISM, ENVIRONMENTALISM, AND THE CULTURE OF UNIQUENESS

The 1968 protest movement (see volume 4), the first of global dimensions, ended at different times in different places in the 1970s, for various reasons. The focus on the means of communication and on the political forces that concentrated above all on the phenomenon of armed terrorist conflict that developed in some Western countries—Germany, and to an even greater extent Italy—

1. Paris, May 1968. The youth demonstrations that took place one after another in the French capital in May of 1968 marked the culmination of the youth protest movement. (Photograph by Ferdinando Scianna/Contrasto.)

2. Worker at the Fiat Mirafiori factory, Turin, Italy, in the 1980s. The young female workers sought emancipation through their work. They brought with them into the factory demands that renewed the practices of the workers' movement. (Photograph by Paola Agosti.)

created a climate of emergency that at times greatly broadened to achieve social control. Less attention was paid to the movements that above all addressed a cultural level that dealt with psychology and daily life. Among these movements, the longest-lived and most important were feminism and environmentalism.

Women, who had been kept at the fringes of the armed conflicts and political and economic events that made the history of the 20th century, began to propose their own distinctive leading role, one that no longer aimed at a simple equality with men but now

included the recognition of gender differences. In other words, women claimed a gender identity that preserved some traditional elements while introducing a strong drive for emancipation.

Environmentalism, the social movement most recently born, asserted itself after it was discovered that progress, nourished by industrial technologies, was reaching limits that could not be violated without triggering uncontrollable ecological crises. Even though strongly opposed by the many interests it threatened, environmentalism held its ground. It fostered a criticism of

3. In an insane asylum: a scene from a theatric performance in Milan, Italy in 1996 inspired by the l of the recluse Adalgisa Conti. One of the most origin aspects of the anti-institutional movement of the 196 and 1970s was its criticism of traditional psychiatry, se as authoritative and repressive. (Photograph by Persili

industrial society that transcended political, cultural, and social strata. Sometimes accused of isolating itself in localism or of defending privileged positions, the movement has assumed global dimensions and fought economic globalization.

The processes of standardization and the loss of memory that are typical of mass culture created a general sameness, even in the most disparate places, and encouraged private consumption. The response to this was a renewed interest in local history, a rediscovery of traditions, cultures, and folkways in small communities. Localism, even more than environmentalism, cannot be traced to a precise political orientation, even if a progressive vision accuses it of being old-fashioned or outmoded. As is true of any point of view that differs from the mainstream, its ambivalence and symptomatic nature are worth emphasizing.

This holds true also for the most surprising and impressive of current movements: the neo-ethnic movement. The resurgence of ethnicity—the search for an identity rooted in heritage and in cultural continuity—appeared, with all its risk of ethnocentrism or even racism, as the other face of the world unification that technology had brought about at the cost of social imbalance and a chaos of worlds and cultures.

4. Canneto sull'Oglio Museum in Lombardy, Italy. The end of the farming culture inspired a recovery of its memory and the founding of local museums. These museums became the expression of the deeper histories of individual communities. (Photograph courtesy of the Museo dell'Oglio e del Chiese.)
5. Avril Quaill, Trespassers Keep Out *(1982). This work underscores the separation between the world of the Aborigines of Australia and the world of the Europeans who menacingly imposed their rules on them.*
6. Mural in San Diego, California. Artistic expression became a means of affirming personal culture and identity. (Photograph by Promotion & Visuals.)
7. Museum of Anthropology in Vancouver, Canada. The ever-increasing speed of modernization, while unifying lifestyles, created a need for the discovery and preservation of the past. This museum became a workshop for the resurgence of native art of the Northwest Coast. (Photograph by Simon Scott.)

11. THE INDIVIDUAL AND THE COLLECTIVE

The great processes of change that characterize the end of the 20th century have not been without consequences to individuals or their choices and aspirations. The sense of having lost, or being on the brink of losing, what remains of local customs and traditions has certainly been one of the effects of globalization. The frailty of collective identity seems no longer to be a temporary crisis, but rather the very shape that development has taken; the same is true of other, more concrete causes for alarm, such as the general reduction in jobs. Neither experience nor professional skill now guarantees economic stability or even survival, and nor do governmental guarantees. It is for this reason that the German scholar Ulrich Beck defines this as a "society of risk," a society in which many feel a need to look after their own interests, to make their own well-being the goal of their every action.

In order to understand this tendency to focus on the needs of one's own family, abandoning any sense of obligation to work toward the common good, we must keep in mind the end of the unification of public will by the old ideologies, the "master narratives" that controlled and motivated millions of people until the very end of the century. Another important factor in this new individualism is the growing disenchantment with political systems that, even when not corrupt or reduced to mere spectacle, are clearly incapable of mediating and of creating consensus. The death of the "master narrative" and this disenchantment with politics both are results of a weakening of government, which in turn is due to the excessive power of multinational enterprises.

The manifestations of the "narcissism

the present—a present incapable of remembering the past or imagining the future. The modern Western sense of the present fails to understand even contemporary situations in other parts of the world; at distances that air travel bridges in a few hours, for example, millions of human beings fight to maintain the most basic living conditions. Mothers, holding their chronically undernourished children, are turned away from aid centers every morning. Myriads of young people undertake a daily harvest on mountains of garbage that surround the shanty-towns in which they live.

What we are witnessing is a general reduction in the opportunities for

6

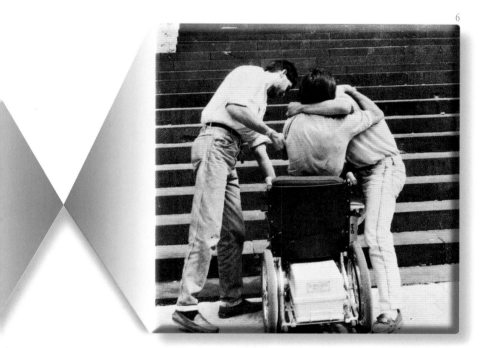

7

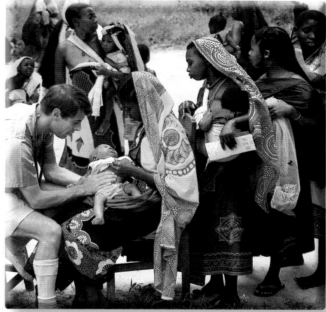

of the masses" that has begun to influence the behavior and expectations of old and young alike, are many and various. The general rejection of social responsibility extends to interpersonal relationships, where there often seems no desire to create long-term bonds of friendship or love. The cult of the body has engendered a technology of well-being that seeks *fitness*, rather than health, aiming to preserve a youthful appearance despite the natural process of aging. This body-consciousness inspires dietary regimens that turn personal weight control into an obsession. The athletic pursuits that it encourages suggest a deeper need to experiment with extreme situations and to experience strong emotions. Also, the loathing of old age, greater possibly than the fear of death, reveals a tendency to live exclusively in and for

interaction, an increasing degradation of meeting places, which become unsafe *because* they are deserted— a clear separation between public life and private life. A dissolution of social interaction appears to follow commerce's penetration of every aspect of

8

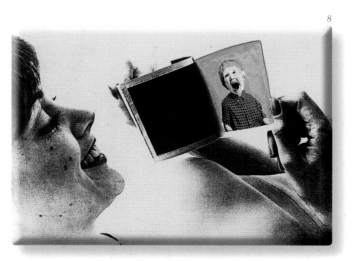

life. Even the elderly, who once could depend on state welfare services, however bureaucratic and impersonal, must now face this change.

Yet, from the perception of a loss of the sense of one's life, from the disenchantment and the sense of bewilderment and solitude that pervade the experience of the inhabitants of industrialized countries, behaviors originate that oppose the ongoing tendencies. The many initiatives of a rising volunteerism have gone beyond the tradition of charity and assistance and proven themselves capable of addressing the very origins of marginalization. They understand both the inadequacy of merely bemoaning the inefficiencies of the system *and* the need to fortify their actions with work toward social justice. Volunteers thus "find time for others" and dedicate themselves to assisting the aged, the handicapped, and the mentally ill; to the rehabilitation of alcoholics, the homeless, people addicted to drugs, and former prisoners; and to education, interculturalism, and the integration of immigrants. They work to foster a sense of global community and a respect for peace; to organize assistance to Third World nations and to fight hunger and war; to create new and egalitarian paths of commerce and a financial system that is ethical and transparent. Among their many larger goals are the care of the environment and a broadening of the accomplishments of culture.

The services of volunteer organizations and other not-for-profit institutions carry considerable economic weight and the have shown themselves capable of creating new social dynamics. In fact, they have formed a "third sector," operating in tandem with governmental and private organizations, that opens areas of global economy and promotes the building of reciprocal relationships—relationships that do not necessarily conform to the logic of a market economy.

The construction of a global economy, the search for ways of life equal to the challenges of the modern world, and the experiences of community life shared by many countries of the West all propose the denial of the closure of individualism and the undertaking of solidarity practices that can fight the crises of stability and security that trouble the developed world.

The accumulation of abandoned cars on the urban periphery is a sign of a means of transportation that, for any number of reasons, its users constantly renew.